Our Universe

Venus

by Margaret J. Goldstein

Lerner Publications Company • Minneapolis

Lerner Publications Company
A division of Lerner Publishing Group
241 First Avenue North
Minneapolis, MN 55401 USA

Website address: www.lernerbooks.com

Words in **bold type** are explained in a glossary on page 30.

Library of Congress Cataloging-in-Publication Data

Goldstein, Margaret J.
Venus / by Margaret J. Goldstein.
p. cm. – (Our universe)
Includes index.
Summary: An introduction to Venus, describing its place in the solar system, its physical characteristics, its movement in space, and other facts about this planet.
ISBN: 0-8225-4649-3 (lib. bdg. : alk. paper)
1. Venus (Planet)–Juvenile literature. [1. Venus (Planet)]
I. Title. II. Series.
QB621 .G66 2003
523.42–dc21 2002000433

Manufactured in the United States of America
1 2 3 4 5 6 – JR – 08 07 06 05 04 03

The photographs in this book are reproduced with permission from: © John Sanford p. 3; NASA, pp. 4, 5, 9, 11, 14, 15, 16, 17, 19, 22, 25, 26, 27; ©Jay Pasachoff/Visuals Unlimited, p. 23.

Cover: NASA.

It is nighttime. Stars and planets shine in the sky. One planet shines the brightest. Which one is it?

The brightest planet in the sky is Venus. Venus is the nearest planet to Earth. Earth is our home planet.

Venus and Earth are part of a group of planets called the **solar system.** There are nine planets in all. The Sun is the center of the solar system.

The closest planet to the Sun is Mercury. Venus is the second planet from the Sun. Earth is the third planet from the Sun.

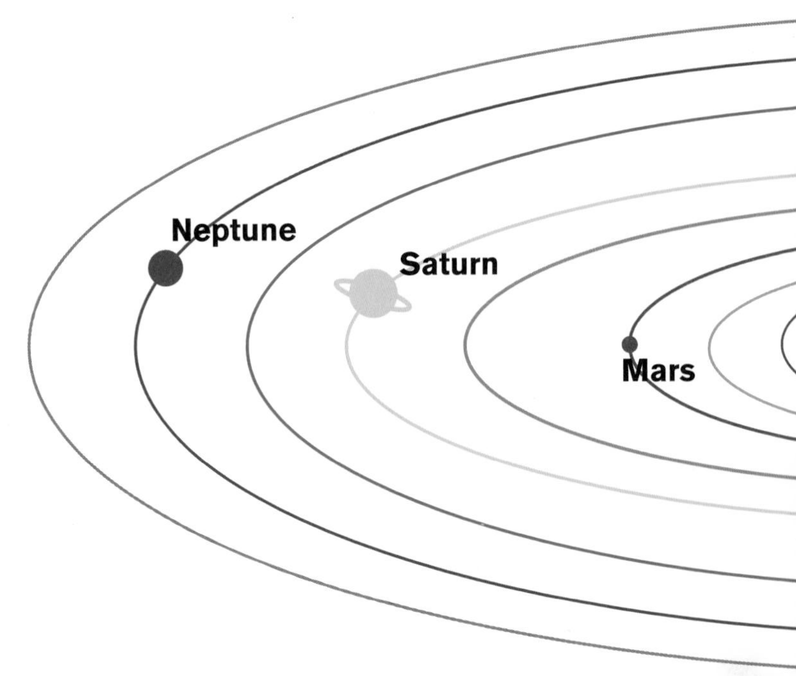

THE SOLAR SYSTEM

All nine planets in the solar system **orbit** the Sun. To orbit the Sun means to travel around it.

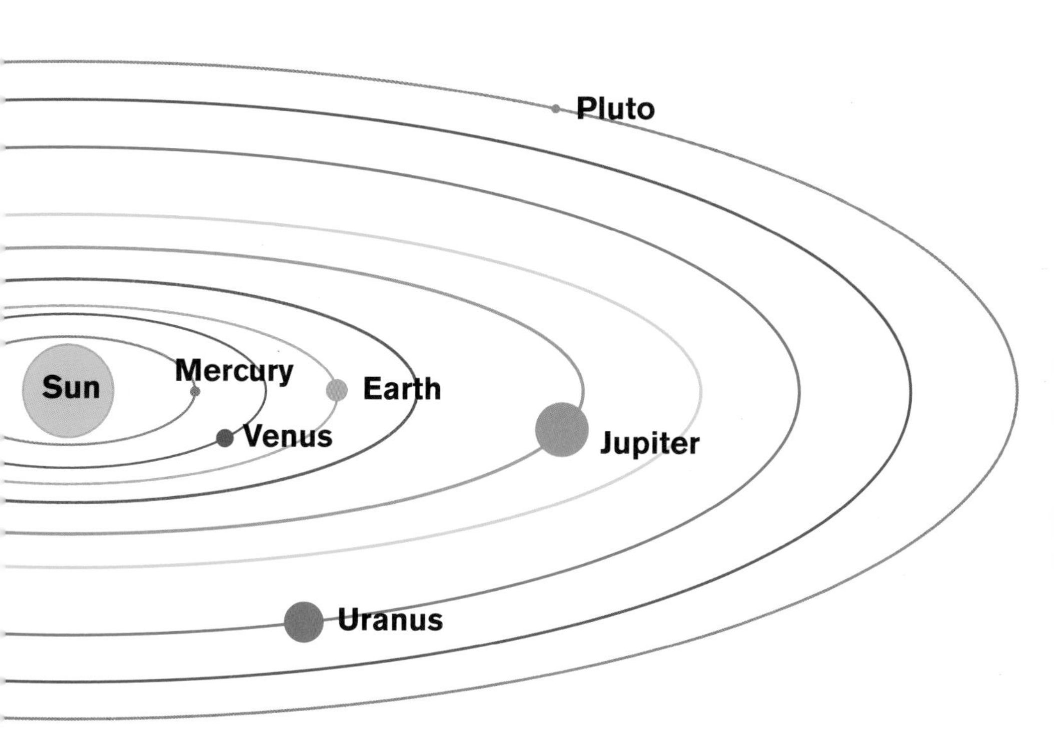

Most of the planets follow oval paths around the Sun. But Venus orbits in nearly a perfect circle. It takes Venus 225 days to orbit the Sun once.

The planets also spin around like tops. This kind of spinning is called **rotating.** How long does it take Venus to rotate?

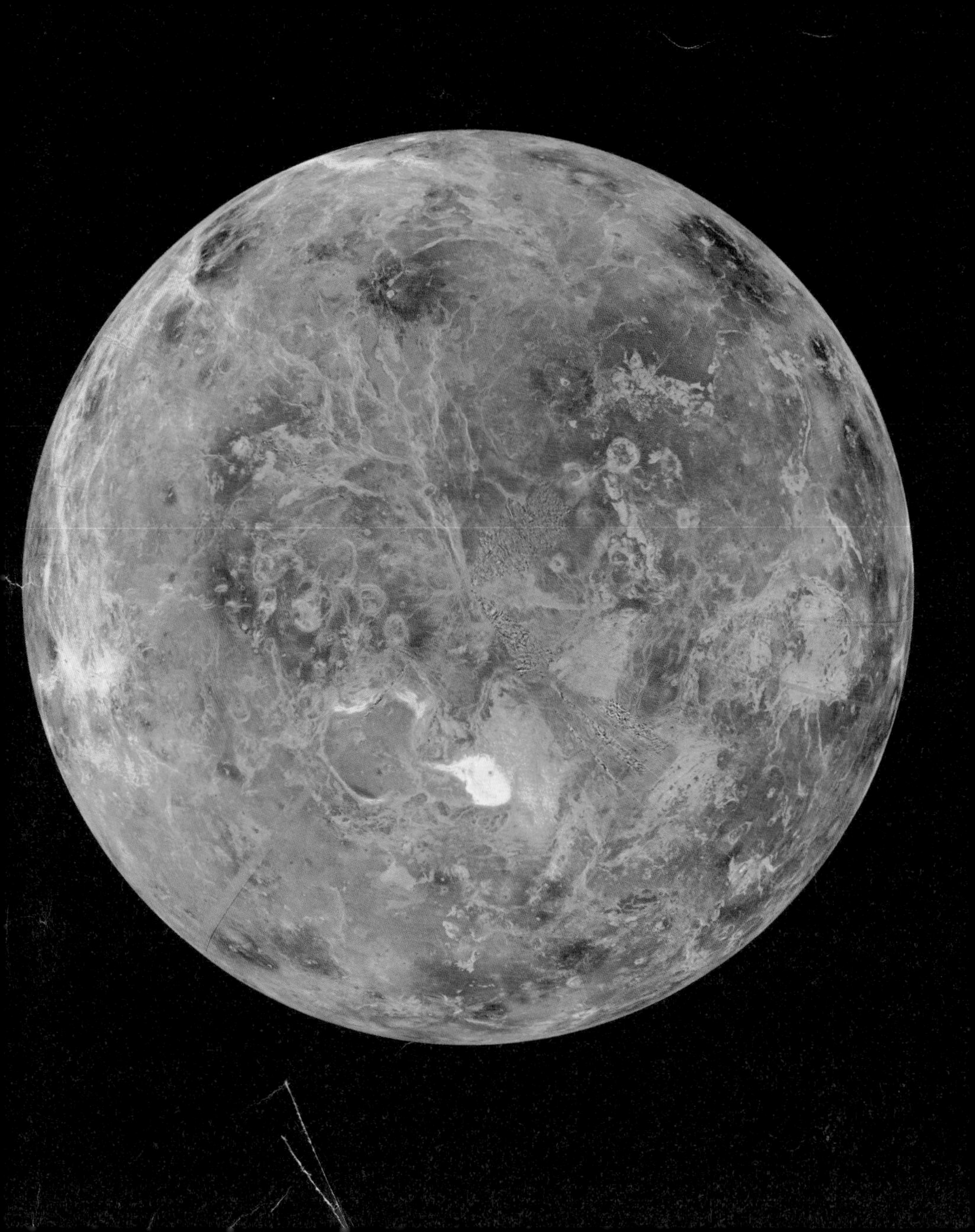

Venus rotates much more slowly than the other planets. It takes 243 days for Venus to spin around once. Earth spins much faster. It rotates in a single day. Venus also rotates backward compared to the other planets in the solar system.

Sometimes Venus is called Earth's twin. The two planets are nearly the same size. Venus is just a little smaller than Earth. How else are the two planets alike?

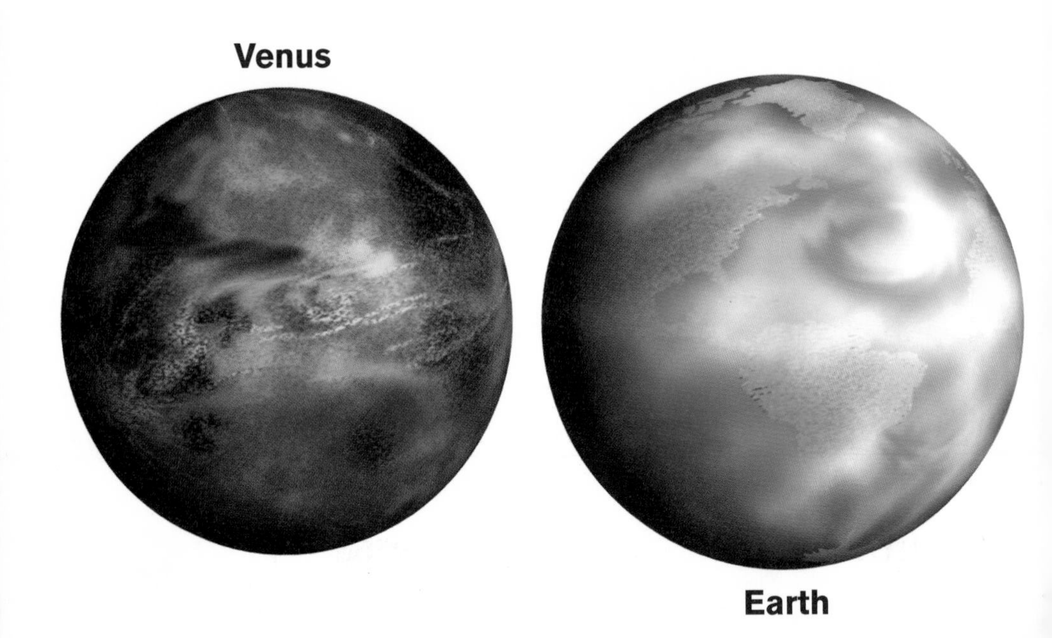

Venus's Layers

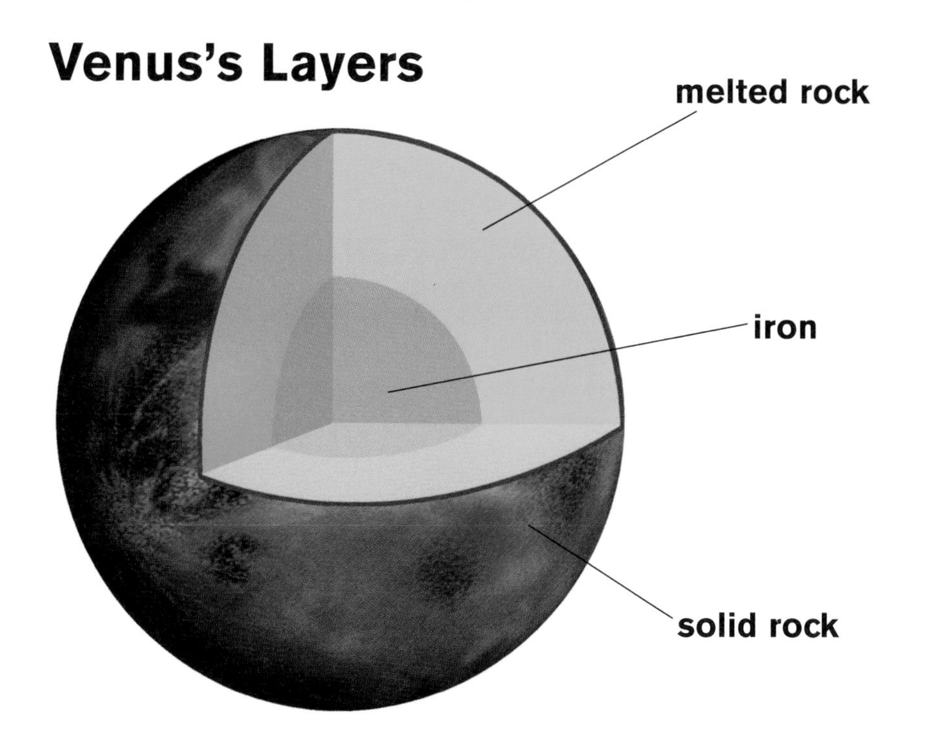

They are both rocky planets. At the center, they are made of a metal called iron. Both planets have a thick layer of melted rock around the center. And both have an outer layer of solid rock.

Thousands of **volcanoes** are found on Venus. Many years ago, hot melted rock burst out of the volcanoes. The melted rock cooled and hardened.

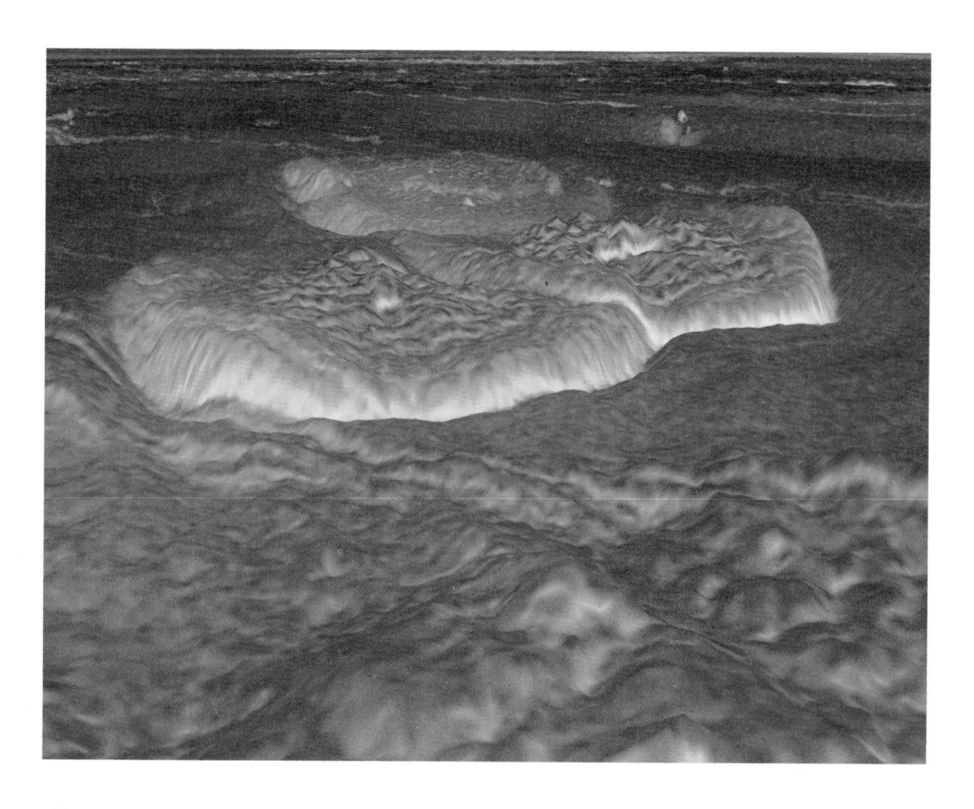

Some of the melted rock hardened to form mountains. Some of it hardened into flat bumps. These bumps are called pancake domes.

Cracks spread out over parts of the ground on Venus. Some of the cracks look like wrinkles. Others look like spiderwebs.

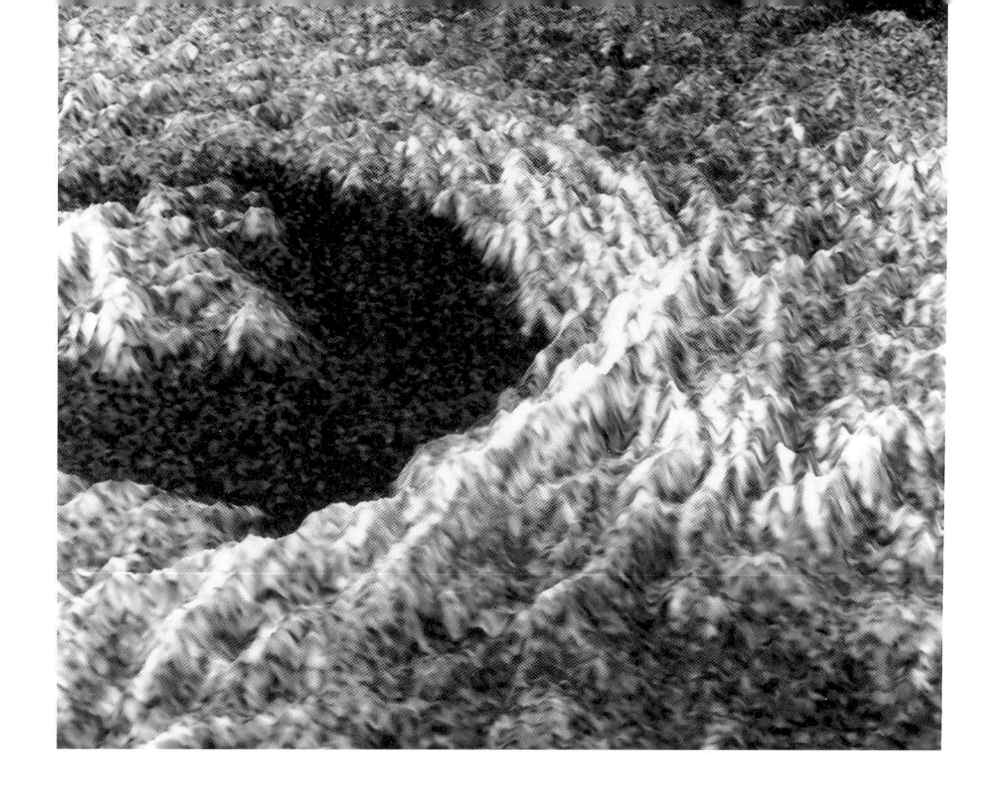

Wide **craters** cover other parts of Venus. Craters are deep holes in the ground. They formed when large pieces of rock from space crashed into Venus.

Venus is surrounded by a thick layer of gases. This layer is called an **atmosphere.** The atmosphere on Venus is made mostly of a gas called carbon dioxide.

The thick atmosphere warms Venus. The atmosphere traps heat from the Sun like windows in a greenhouse. This warming of the planet is called the **greenhouse effect.**

The greenhouse effect makes Venus burning hot. Temperatures on the planet are hundreds of degrees above zero. That is much hotter than any place on Earth.

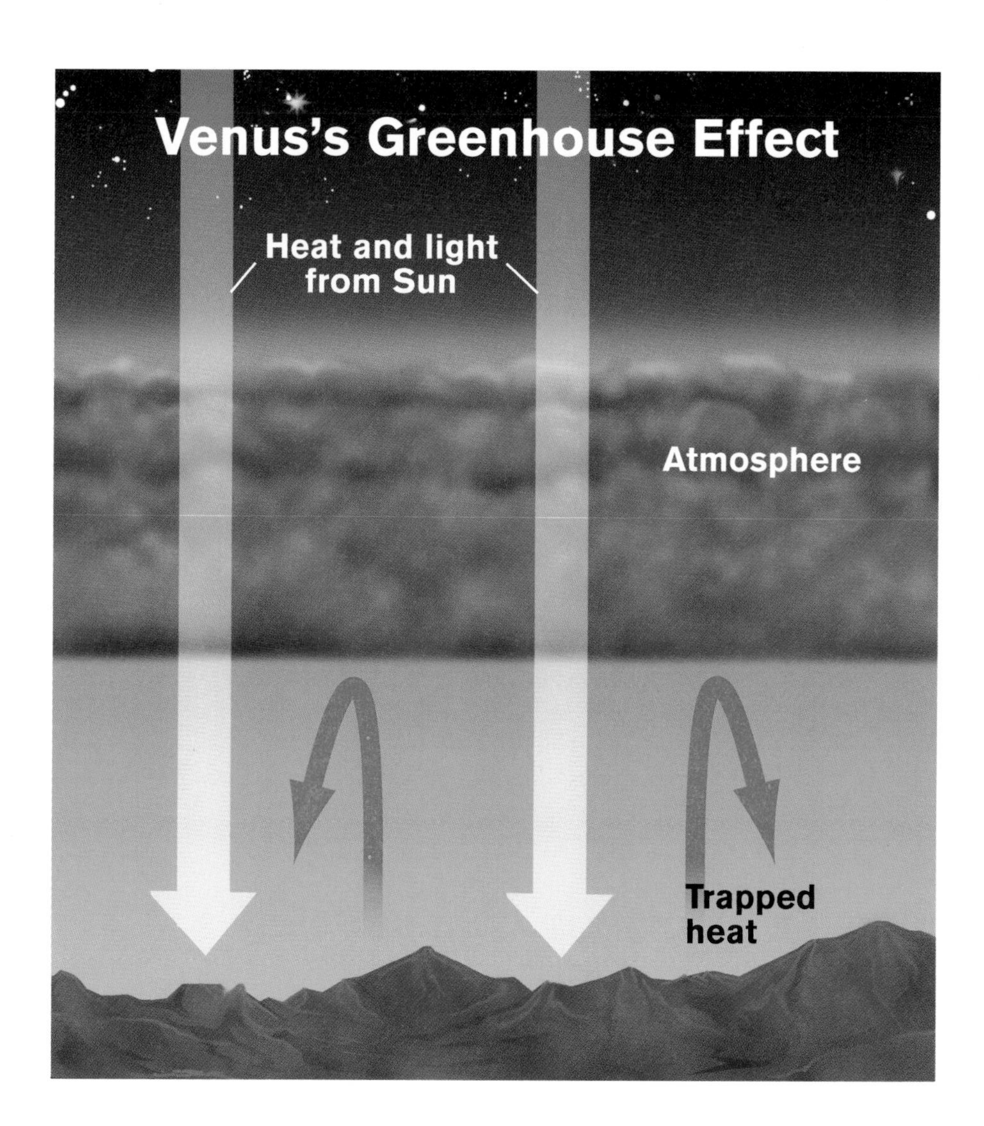
Venus's Greenhouse Effect
Heat and light
from Sun
Atmosphere
Trapped
heat

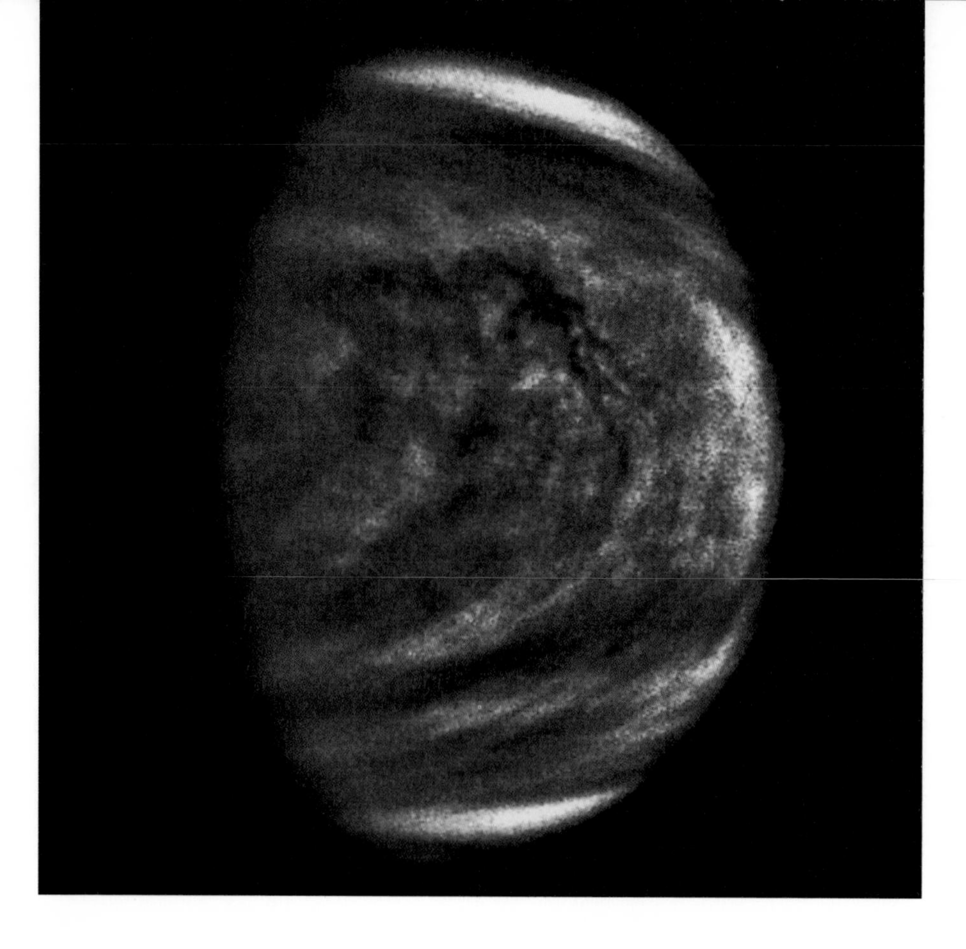

Many thick clouds float in Venus's atmosphere. Strong winds blow the clouds across the sky at great speeds.

Light from the Sun bounces off the tops of the clouds. The sunlight makes Venus shine brightly.

Starting in the 1960s, people sent many spacecraft to visit Venus. They hoped to learn more about the planet.

The spacecraft carried machines and cameras. The machines studied Venus's atmosphere, clouds, and land. The cameras took many photographs.

USA

An artist made this picture of a spacecraft on its way to Venus.

About 20 spacecraft have visited Venus. They have taught us a lot about the planet. But we have much more to learn.

Look for Venus in the sky. What else would you like to know about the closest planet to Earth?

Facts about Venus

- Venus is 67,200,000 miles (108,000,000 km) from the Sun.
- Venus's diameter (distance across) is 7,520 miles (12,100 km).
- Venus orbits the Sun in 225 days.
- Venus rotates in 243 days.
- The average temperature on Venus is 890°F (480°C).
- Venus's atmosphere is made of carbon dioxide and nitrogen.
- Venus has no moons.
- Venus was named after the Roman goddess of love and beauty.
- Venus has been visited by the American spacecraft *Mariner 2* in 1962–63,

Mariner 5 in 1967, *Mariner 10* in 1973–75, *Pioneer Venus* in 1978, and *Magellan* in 1989–1994.

- Because of Venus's unusual rotation, a person standing on Venus would see the Sun rise in the west and set in the east.
- Venus may have once had oceans and rivers.
- Venus is nicknamed the morning star and the evening star because it is often seen at these times of day.
- Venus was the first planet to be visited by a spacecraft.
- Twenty space missions have visited Venus. Five missions came from the United States. Fifteen came from the former Soviet Union.

Glossary

atmosphere: the layer of gases that surrounds a planet or moon

craters: large holes on a planet or moon

greenhouse effect: the warming of a planet, caused when its atmosphere traps heat from the Sun

orbit: to travel around a larger body in space

rotating: spinning around like a top

solar system: the Sun and the planets, moons, and other objects that travel around it

volcanoes: openings in the surface of a planet or moon. Hot rock and gases sometimes burst up through the openings.

Learn More about Venus

Books

Brimner, Larry Dane. *Venus.* New York: Children's Press, 1999.

Simon, Seymour. *Venus.* New York: Morrow, 1998.

Websites

Solar System Exploration: Venus
<http://solarsystem.nasa.gov/features/planets/venus/venus.html>
Detailed information from the National Aeronautics and Space Administration (NASA) about Venus, with good links to other helpful websites.

The Space Place
<http://spaceplace.jpl.nasa.gov>
An astronomy website for kids developed by NASA's Jet Propulsion Laboratory.

StarChild
<http://starchild.gsfc.nasa.gov/docs/StarChild/StarChild.html>
An online learning center for young astronomers, sponsored by NASA.

Index